Instant PLC Programming with RSLogix 5000

Learn how to create PLC programs using RSLogix 5000 and the industry's best practices using simple, hands-on recipes

Austin Scott

BIRMINGHAM - MUMBAI

Instant PLC Programming with RSLogix 5000

First published: October 2013

Production Reference: 1251013

Published by Packt Publishing Ltd.
Livery Place
35 Livery Street
Birmingham B3 2PB, UK.

ISBN 978-1-84969-844-3

www.packtpub.com

Credits

Author

Austin Scott

Reviewers

Prof. Vinod B. Kumbhar

G. Scott Whitlock, P.Eng.

Acquisition Editors

Aarthi Kumaraswamy

Rubal Kaur

Commissioning Editor

Govindan K

Technical Editors

Anita Nayak

Pratik More

Copy Editor

Tanvi Gaitonde

Project Coordinator

Esha Thakker

Proofreader

Jonathan Todd

Production Coordinator

Pooja Chiplunkar

Cover Work

Alwin Roy

Cover Image

Yuvraj Mannari

About the Author

Austin Scott, in 2006, founded Synergist SCADA, a successful company that provided vendor-neutral SCADA architecture and development. Synergist has also developed a suite of engineering tools, including Citect Power Tools and Active Network Security. In July 2013, Synergist was acquired by Cimation LLC as the catalyst for its growing Canadian operations and ongoing product development.

With more than a decade of industrial automation and software development experience, Austin has worked on large-scale, high-profile projects across North America and globally, incorporating most major SCADA platforms.

Austin's professional focus includes developing and refining custom software solutions to enhance the productivity of SCADA developers, improve integration between SCADA data and corporate applications, and cyber security, especially the detection of unauthorized access to SCADA networks and forensic analysis of SCADA breaches.

Cimation and its 250 talented employees serve North America's energy industry with automation and controls, industrial IT, and enterprise data solutions (inc. SCADA).

Forbes recently ranked Cimation number 22 on its second annual America's Most Promising Companies list. Cimation was the only energy consulting firm ranked in the list's top 25. The Forbes America's Most Promising Companies list is comprised of 100 privately held, high-growth companies with bright futures. Most of the companies ranked on this year's list, Cimation included, fall into the technology category. Since its inception in 2009, Cimation has grown 1,397 percent in revenue.

Cimation is headquartered in Houston with regional offices in Calgary, Denver, Pittsburgh, New Orleans, and Lafayette.

About the Reviewers

Prof. Vinod B. Kumbhar is an Assistant Professor at Adarsh Institute of Technology & Research Centre (AITRC), Shivaji University, India specializing in developments in Industrial Automation Technology. Prior to joining AITRC, he worked as a Senior Customer Support Engineer at Messung Systems Pvt.Ltd, a leading automation company in India.

He has executed a good number of turn-key automation systems, from designing, programming, and developing, to commissioning in industrial automation. He has worked with various brands of PLCs such as Siemens, Allen-Bradley, Mitsubishi, and ABB. Currently he is involved in research work on communications between PLC and Visual Basic for cost-effective SCADA solutions.

He is currently pursuing an M.E. in Electronics & Telecommunication from Shivaji University. He has published several papers on PLC-based systems. He is the owner of the blog `http://plc-scada-dcs.blogspot.com`, where he writes tutorials on the latest developments in automation engineering.

Scott Whitlock is a licensed professional engineer and automation programmer. He has been writing programs in RSLogix 500 and 5000 since graduating from the University of Waterloo in 2000. He is the author of a popular (and free) online tutorial for RSLogix 5000 beginners. He also does a lot of HMI and data-collection work using C# and SQL Server, respectively. Recently he has been building more systems using TwinCAT 3, a PC-based control system. Outside of work, he has written and released two open source libraries (called FluentDwelling and SoapBox Core) and one open source ladder logic editor (called SoapBox Snap), all written in C#. Scott still lives in the small town where he grew up, with his loving wife Tammy, three adorable young children, and their family dog, Roxy. In those precious minutes after the kids go to bed, Scott sometimes finds the time to write down his thoughts about PLC programming, PC programming, and the differences between the two on his blog, *Contact and Coil*, at `http://www.contactandcoil.com/`.

www.PacktPub.com

Support files, eBooks, discount offers, and more

You might want to visit www.PacktPub.com for support files and downloads related to your book.

Did you know that Packt Publishing offers eBook versions of every book published, with PDF and ePub files available? You can upgrade to the eBook version at www.PacktPub.com and as a print book customer, you are entitled to a discount on the eBook copy. Get in touch with us at service@packtpub.com for more details.

At www.PacktPub.com, you can also read a collection of free technical articles, sign up for a range of free newsletters, and receive exclusive discounts and offers on Packt Publishing books and eBooks.

http://PacktLib.PacktPub.com

Do you need instant solutions to your IT questions? PacktLib is Packt Publishing's online digital book library. Here, you can access, read, and search across Packt Publishing's entire library of books.

Why Subscribe?

- Fully searchable across every book published by Packt
- Copy and paste, print, and bookmark content
- On demand and accessible via web browser

Free Access for Packt account holders

If you have an account with Packt at www.PacktPub.com, you can use this to access PacktLib today and view nine entirely free books. Simply use your login credentials for immediate access.

For my traveling buddy, best friend, and loving wife, Maljori, who has always supported me in the pursuit of my dreams.

Austin Scott

Table of Contents

Preface

The world of industrial automation is witnessing an unprecedented change, much like the IT world has experienced over the past decade. Automation professionals must upgrade their skills constantly with changing technology. There is rarely any time or budget for a week long training course on new platforms. Skilled professionals learn by themselves, making use of resources that are readily available within their organization and on the Internet. In the content of this book, the fundamentals for each key feature of RSLogix 5000 are presented along with links to online resources. The ultimate goal is to provide the reader with as much detail on the RSLogix 5000 platform as required. This short book provides a modern automation professional with all the information needed to become an expert in North America's most popular PLC platform.

What this book covers

Creating a new RSLogix 5000 application (Simple) steps through the initial setup of a new project, rack, and processor. We also introduce RSLogix Emulate 5000, which can be configured in place of a physical PLC processor and rack. This recipe explains the basic setup of a simulated PLC processor and module configuration.

Configuring the I/O modules (Simple) introduces us to the process of adding and configuring I/O modules in RSLogix 5000. We cover the procedure for configuring the signal range and scaling it to an engineering range. We also touch on module alarm configuration and installing third-party modules.

Understanding tags (Simple) provides an overview of the various data types that are available in RSLogix 5000's text-based tagging system, tag scope, and how to create new tags using a few different methods. We also introduce the user to unique tag types, produced and consumed tags, which can be used for controller-to-controller communication.

Building Ladder Diagram programs (Simple) takes a first look at creating new routines using ladder logic diagrams. The reader is introduced to the concept of Tasks and also learns how to link routines. In this recipe, we learn how to navigate the ladder elements that are available, how to find help on each element, and how to create a simple alarm timer using ladder logic.

Troubleshooting techniques (Intermediate) are essential when things go awry. In this recipe, we take a look at a few of the basic techniques that will aid us in resolving issues with our program and the PLC.

Downloading explained (Simple) demonstrates how to load and run your application on a PLC in a step-by-step process. We also warn about some of the dangers of downloading to a running plant.

Uploading explained (Simple) takes us through the procedure of loading an existing application from a running PLC to RSLogix 5000 in a step-by-step process. Common pitfalls and problems are also discussed in this recipe.

Understanding online changes (Intermediate) will introduce the concept of online changes in RSLogix 5000 and provide insights into the limitations you will encounter. In this recipe, we make a change to a PLC program without downloading the PLC again or stopping the execution.

Building Functional Block Diagrams (Simple) demonstrates how to create a new FBD routine to handle our valve alarms. We also learn about the advantages of using FBD properties, how they easily integrate with an HMI, and explore the limitations you may run into when using FBDs.

Building a Structured Text program (Simple) sets up a calculation for our project and outlines the best practices and limitations of using structured text.

Building a Sequential Function Chart (Intermediate) provides a detailed illustration of how an SFC can be used to create a backwash sequence; our example explains the key components that make up an SPC. We demonstrate how structured text is used within SFC steps and the limitations you may encounter.

Organizing your project code (Advanced) explores the way RSLogix organizes its controller into tasks, programs, and routines. Also, we will gain a deeper understanding of tag scope and how it can impact our routines.

Exporting tags (Simple) shows us a handy way to generate reports on your RSLogix 5000 application and create a tag list for integration with an HMI. We also explore the ability to import tags after making modifications to them and some of the potential problems that you can run into when doing this.

Exporting programs (Simple) will demonstrate how to export a program to an XML (Extensible Markup Language) file for manipulation or reporting. The ability to import modified XML files into our programs is also explored.

Printing programs (Simple) introduces us to the powerful printing options that are available in RSLogix 5000. We view the custom options available with each routine type and the routine types that can benefit from a larger page sheet size.

Code generation (Advanced) shows us a handy trick in RSLogix 5000—how the same logic can be replicated many times but with different variable tags. This can easily be accomplished in Logix5000 using Notepad and a simple code-generation trick (of the true Logix 5000 masters).

Understanding user-defined data types (Advanced) introduces a structured, easily maintainable, and highly repeatable method for building ladder logic. We demonstrate how UDTs can streamline our program and touch on advanced topics such as nesting and the types of data that can be used.

What you need for this book

In order to complete the chapters in this book, you will need RSLogix 5000 Version 17 or higher. Access to a Logix 5000 PLC is also recommended, but Emulate 5000 can also be used.

Who this book is for

The purpose of this book is to capture the core elements of PLC programming with RSLogix 5000 so that electricians, instrumentation techs, automation professionals, and students who are familiar with basic PLC programming techniques can come up to speed with minimal investment of time and energy. I have intentionally avoided getting into any background information about control theory or IEC61131-3, and have focused on condensing information specific to RSLogix 5000. I have selected the key areas that separate RSLogix 5000 from other programming environments and have provided a step-by-step, cookbook approach to learning them.

Conventions

In this book, you will find a number of styles of text that distinguish between different kinds of information. Here are some examples of these styles, and an explanation of their meaning.

Code words in text are shown as follows: "Type in the structured text logical statement `FC1001_PV=100`."

New terms and **important words** are shown in bold. Words that you see on the screen, in menus or dialog boxes for example, appear in the text like this: "Click on the **OK** button to save the new tag."

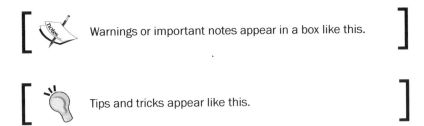

Warnings or important notes appear in a box like this.

Tips and tricks appear like this.

Reader feedback

Feedback from our readers is always welcome. Let us know what you think about this book—what you liked or may have disliked. Reader feedback is important for us to develop titles that you really get the most out of.

To send us general feedback, simply send an e-mail to feedback@packtpub.com, and mention the book title via the subject of your message.

If there is a topic that you have expertise in and you are interested in either writing or contributing to a book, see our author guide on www.packtpub.com/authors.

Customer support

Now that you are the proud owner of a Packt book, we have a number of things to help you to get the most from your purchase.

Errata

Although we have taken every care to ensure the accuracy of our content, mistakes do happen. If you find a mistake in one of our books—maybe a mistake in the text or the code—we would be grateful if you would report this to us. By doing so, you can save other readers from frustration and help us improve subsequent versions of this book. If you find any errata, please report them by visiting http://www.packtpub.com/submit-errata, selecting your book, clicking on the **errata submission form** link, and entering the details of your errata. Once your errata are verified, your submission will be accepted and the errata will be uploaded on our website, or added to any list of existing errata, under the Errata section of that title. Any existing errata can be viewed by selecting your title from http://www.packtpub.com/support.

Piracy

Piracy of copyright material on the Internet is an ongoing problem across all media. At Packt, we take the protection of our copyright and licenses very seriously. If you come across any illegal copies of our works, in any form, on the Internet, please provide us with the location address or website name immediately so that we can pursue a remedy.

Please contact us at copyright@packtpub.com with a link to the suspected pirated material.

We appreciate your help in protecting our authors, and our ability to bring you valuable content.

Questions

You can contact us at questions@packtpub.com if you are having a problem with any aspect of the book, and we will do our best to address it.

Instant PLC Programming with RSLogix 5000

Welcome to *Instant PLC Programming with RSLogix 5000*. The purpose of this book is to capture the core elements of PLC programming with RSLogix 5000 so that automation professionals or students who are familiar with other PLC programming environments can come up to speed with minimal investment of time and energy.

About RSLogix 5000

Rockwell Automation and its sister company Rockwell Collins are part of an impressive legacy, including the space shuttle program, the Apollo moon landings, P-51 mustang, the Navstar **Global Positioning System** (**GPS**) satellites, and the supersonic B-1 Lancer. Today, it is a global provider of industrial automation, power, control, and information solutions under the brands Allen-Bradley and Rockwell Software. Rockwell Automation products are extremely popular in the North American market and there are references to it being used in almost every industry vertical.

RSLogix 5000 is a user-friendly, **IEC61131-3**-compliant interface for programming the current generation of Rockwell Automation **PLC**s (**Programmable Logic Controllers**). IEC61131-3-compliant means that it complies with the International Open Standard for PLC programming languages, including:

- ▸ **Ladder Diagram** (**LD**)
- ▸ **Function Block Diagram** (**FBD**)
- ▸ **Structured Text** (**ST**)
- ▸ **Sequential Function Chart** (**SFC**)

More information on RSLogix 5000 IEC compliance can be found in the Rockwell publication *Logix5000 Controllers IEC 61131-3 Compliance* available at `http://literature.rockwellautomation.com/idc/groups/literature/documents/pm/1756-pm018_-en-p.pdf`.

The RSLogix 5000 platform is compatible with the new generation of Allen-Bradley controllers, including ControlLogix, CompactLogix, FlexLogix, SoftLogix, RSLogix Emulate 5000, and GuardLogix.

RSLogix 5000 is one of several PLC programming environments from Allen-Bradley, so it is critical to work with Rockwell Automation sales and support staff to ensure that you are using compatible software, firmware, and hardware. Throughout the book, we have used RSLogix 5000 Version 19; however, the same principles apply to older and newer versions (Version 20) of the software.

Installing RSLogix 5000 (Simple)

RSLogix 5000 can be purchased from your local Rockwell distributor and installed from the CD provided. There is a demo (Version 17) of RSLogix 5000 that can be downloaded from the following URL:

`http://www.rockwellautomation.com/rockwellsoftware/design/rslogix5000/demo.html`

The recommended best practice for installing RSLogix 5000 is to follow the official installation notes by Rockwell, which can be downloaded from the following URL:

`https://rockwellautomation.custhelp.com/ci/fattach/get/176048/`

Creating a new RSLogix 5000 application (Simple)

In this recipe, we will create a new RSLogix 5000 controller program using the Logix Designer application.

Getting ready

For this exercise, you will need to have RSLogix 5000 installed on your computer.

How to do it...

1. First, we will need to open RSLogix 5000 from the drop-down menu and then navigate to **File | New** or press *Ctrl + N*.

2. The **New Project** window will appear and allow us to give a name to our new project and configure it for a particular controller and firmware revision. In this book, we will be using the 1756-L75 with firmware Version 19. Feel free to use whichever hardware is available or the RSLogix Emulate 5000 Controller if a PLC is not available; but, keep in mind that the hardware may impact your ability to complete all the exercises in this book.

> The general rule of thumb is that the major version numbers must always match (so, for Version 20.x of RSLogix 5000, you need Version 20.x of the controller firmware). It is common practice to have multiple RSLogix software versions installed side by side on a computer, particularly when you're dealing with a plant that has older controllers that have not been upgraded.

3. Set the remaining fields to the following values:

   - **Name**: FirstProject
   - **Description**: This is my first project in RSLogix 5000!
   - **Chassis Type**: 4-Slot
   - **Slot**: 0

4. Click on the **Browse** button to choose a location on your computer on which to save your project.

How it works...

We have created a new empty RSLogix 5000 project and specified the PLC that will be used throughout this book. When you create a new project, it will be stored on your local computer as an ACD file.

There's more...

If you have selected the RSLogix Emulate 5000 controller type, you will also need to configure it using the RSLogix Emulate 5000 program separately to simulate a PLC on your local computer.

RSLogix Emulate 5000

RSLogix Emulate 5000 allows you to use your local computer as a Logix 5000 PLC. You will need to have RSLogix Emulate 5000 running and configured in order to test the projects described in this book. RSLogix Emulate 5000 is very simple to set up; simply run the RSLogix Emulate 5000 program and switch between the **Run** mode and the **Program** mode using the **All Modules** drop-down menu. For the purposes of this book, there is no need to configure any modules.

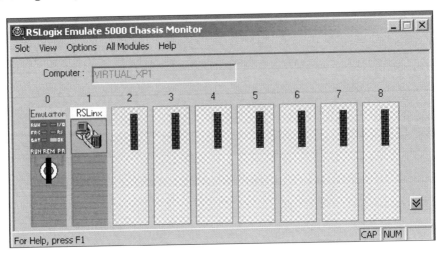

Allen-Bradley Rockwell programmable controllers

There is a wide range of programmable controllers available from Rockwell—too many than can be mentioned in this book. It is worth spending some time learning the full line-up of controllers on its website at the following link:

http://ab.rockwellautomation.com/Programmable-Controllers

Configuring I/O modules (Simple)

In this recipe, we are going to configure an analog input card to monitor the open position of 4 valves (0 to 100 percent) based on their 4 to 20 ma signals.

Getting ready

In order to complete this exercise, you will need a Logix 5000-compatible rack, a PLC, and some compatible controller cards, or run RSLogix Emulate 5000.

How to do it...

1. At the bottom of the **Controller Organizer** window, expand the **I/O Configuration** folder, right-click on the **Backplane** node, and select **New Module...** from the menu as shown in the following screenshot:

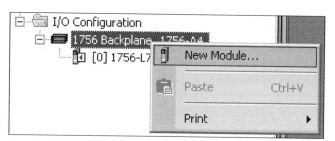

2. Next, we will select the module that will be added to our rack. We will be adding a basic 8 channel analog input card. Expand the **Allen Bradley Analog** category and choose **1756-IF8**.

3. Now, we are presented with the **New Module** window where you can specify the name and description of the module. We will identify this module based on its rack and slot position—RACK01_SLOT01 as shown in the following screenshot:

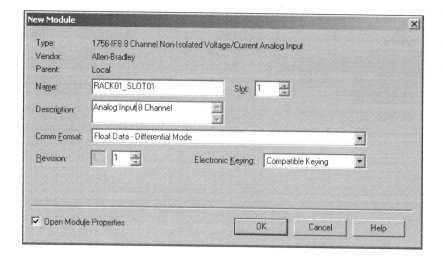

 In our example, we use an easy-to-understand module name, RACK01_SLOT01. In a real-world environment where you will be referencing the module name frequently, you may want to use a shorter naming convention, like R01S01.

4. After adding the new module, the **Module Properties** window is displayed (or can be opened by double-clicking on the module). The **Module Properties** window will allow us to configure the channel, calibrate the card, and modify the **Requested Packet Interval** (**RPI**) and alarm information of our analog input module.

The RPI is the millisecond interval (0.2 ms to 750 ms) during which the CPU will request new I/O states from your I/O cards. The ControlLogix line of PLCs contains multiple CPU cores and is capable of executing multiple tasks simultaneously. Unlike previous generations of PLCs, the ControlLogix will scan its PLC logic and update I/O at the same time (asynchronous I/O updates). Therefore, it is possible for I/O values to change in the middle of your logic scan. More information on the way ControlLogix executes tasks can be found in the Rockwell publication *Logix5000 Controllers Tasks, Programs, and Routines* available at `http://literature.rockwellautomation.com/idc/groups/literature/documents/pm/1756-pm005_-en-p.pdf`.

5. We will configure our analog input module's first channel (**Channel 0**) to be a typical 4 ma to 20 ma input by navigating to the **Configuration** tab and selecting an input range of **0 ma to 20 ma** from the dropdown.

6. Now, we will set the information in the **Scaling** group box to the following values:

    ❑ **High Signal**: 20 ma

    ❑ **Low Signal**: 4 ma

    ❑ **High Engineering**: 100.0

    ❑ **Low Engineering**: 0.0

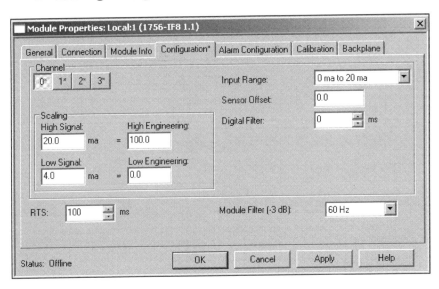

7. This process must be repeated for all the four channels. We can select and configure the three remaining channels by clicking on the small square buttons labeled **1**, **2**, **3**, and following the same method.

8. Finally, we will click on the **OK** button to save our module configuration.

How it works...

We have configured all four channels on our module to report our valve open positions from 0 to 100 based on a 4 ma to 20 ma signal. If a valve signal wire is disconnected, we will see that our channel will drop below 4 ma and a channel fault will be reported.

There's more...

Each card type will have its own configuration dialog, so take some time to learn the available cards and their configuration options.

Third-party modules

Logix 5000 supports a wide variety of third-party add-on modules. Most of these modules will need to be installed using an installer from the third-party company; however, in Version 20 of Logix 5000, many popular third-party modules are built into the product.

Understanding tags (Simple)

RSLogix 5000 **Named Associations** (text-based memory locations also known as variables) are used in our PLC application to represent process values and equipment. In this recipe, we will introduce text-based name associations used in RSLogix 5000. We will demonstrate the configuration of tag types **Base** and **Alias**. We will work through a step-by-step guide to creating and associating tags with inputs, outputs, and memory.

Getting ready

To complete this exercise, you need to have completed the previous recipes.

How to do it...

1. First, we will create a tag using the **New Tag** window. You can open the new tag window by right-clicking on the **Controller Tags** folder of the **Controller Organizer** (to display the **Controller Organizer**, press *Alt + O*) and selecting **New Tag** (or by pressing *Ctrl + W*).

2. The **New Tag** window allows us to configure several parameters for our tag. The first tag we will create will be a base tag that will be used internally in the program and is not directly associated to a card in the controller. The first base tag will be the set point for the flow control valve FC1001, so we will give it a text-based named association of FC1001_SP.

3. Along with the name, setting a description is a recommended best practice. We will set the description to FLOW CONTROL 1001 SET POINT. Refer to the upcoming screenshot.

4. Next, we will set the **Type** dropdown to **Base**.

5. We will set the **Data Type** of our base tag to REAL.

 The REAL data type in RSLogix 5000 is a 32-bit (4 byte) value based on the IEEE 754 single-precision format. Using a floating decimal point, it is capable of representing a wide, dynamic range of values at the cost of precision. Data types are a critical component of PLC programming with the RSLogix 5000 platform. There are dozens of data types available in RSLogix 5000—too many to list in this short book. Rockwell has published a terrific document, *Logix 5000 Controllers I/O and Tag Data*, detailing the available data types and their usages available at http://literature.rockwellautomation.com/idc/groups/literature/documents/pm/1756-pm004_-en-p.pdf.

6. We will set the value of **Scope** to **FirstController**.

7. Setting the scope to **Controller level** will allow us to use this tag in all the tasks and programs in the project (Global scope).

8. The **External Access** field is set to **Read/Write** so that our operator will be able to set the value from the HMI.

9. The **Style** dropdown allows us to modify the way the value is displayed and the choices vary depending on the data type that is being used. The following screenshot shows all of these settings:

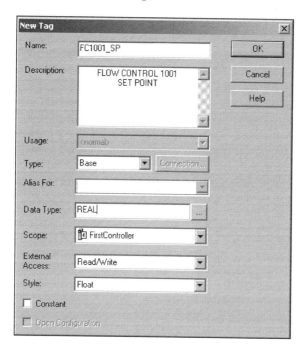

10. Finally, we will click on **OK** to finish adding our variable tag to the project.

11. Now, we will double-click on the **Controller Tags** leaf of the **Controller Organizer** and view our newly added tag in the **Controller Scope** pane of our project.

The **Tag Properties** window can be viewed by selecting the tag, then right-clicking on it and selecting **Tag Properties** (or pressing *Alt + Enter*).

12. Next, we will repeat the same process for an **Alias** tag that will be directly associated with the analog output of the flow control valve. This time, we will use the **Controller Tags** pane to enter our new tag into the **Tags** table. At the bottom of the **Controller Tags** pane, there is a tab labeled **Edit Tags** and, if we click on that tab, we can type a tag into the bottom row of the table as if we were adding it to a spreadsheet.

13. In the **Name** column, set the value to FC1001_PV.

14. In the **Alias for** column, select the data outputted from the analog input card. We added `Local:1:I.Ch0Data` as shown in the following screenshot:

FC1001_SP			REAL
FC1001_PV	Local:1:I.Ch0Data	Local:1:I.Ch0Data	REAL

15. Finally, set the description to `FLOW CONTROL VALVE 1001 PROCESS VALUE`.

Using aliases for inputs is a bad practice due to the asynchronous nature of the way the ControlLogix processor requests data. The ControlLogix CPU has multiple cores capable of executing multiple tasks simultaneously; your analog input values can change part way through the scan of your logic. Values changing part way through a routine can cause results that are difficult to predict and put a program in a state that it is not intended to be in. Experienced programmers copy the value of the input to an internal tag at the beginning of the routine or at the beginning of the continuous task. Using COP instructions to move digital input values to internal tags is also a bad practice because it is more difficult to force the values. Forcing values that are mapped contact-to-coil is much easier to do and read. In summary, digital inputs should be mapped coil-to-contact to internal tags and analog inputs should be mapped to internal tags using MOV instructions.

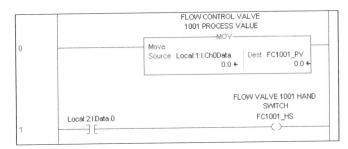

Outputs use aliases because there are no issues with asynchronous data transfers and it is easy to force an alias.

How it works...

We have added two new tags to our project as follows:

► A base type tag that is not directly connected to an input or output card and allows us to specify any data type we wish to use

► An alias type tag that is connected to an input or output card value and whose data type is specified by the card configuration

RSLogix 5000 uses text-based named associations that can be referenced by name throughout the project. In RSLogix, tags can also be referenced by external sources, such as HMIs, SCADA, and DCS systems rather than the numeric addresses of legacy PLC applications.

There's more...

Text-based named associations is an important concept in RSLogix 5000. This recipe describes a few more important aspects of it.

Deleting tags

Tags can also be deleted from the **Edit Tags** tab of the **Controller Tags** pane. You can delete a tag by right-clicking on the box to the left of the name and selecting the **Delete** menu option (or by pressing the *Delete* key after selecting a tag).

Understanding tag scope

The **Scope** field of the **New Tag** form represents the area where the tag is visible within the controller. When you specify the scope as the controller itself (**FirstController** in our example), the tag is accessible globally, meaning it can be accessed from all tasks and programs. Later, we will create programs and define tags that have a scope for a particular program and cannot be accessed outside that specific program.

Data Types in RSLogix 5000

There is a wide range of data types available, which can be seen from the **Data Type** dropdown of the **New Tag** form. Base types, arrays, function block types, motion control types, and so on can be added to a controller at any **Scope** level.

Produced and Consumed tags

You may have noticed that there were two other options in the **Type** field of the **New Tag** form: **Produced** and **Consumed**. These tags allow values to be shared between two separate PLCs on a network. They are limited to a handful of data types. The project that contains the **Produced** tag must have the remote-networked PLC mapped in **I/O Configuration** of the project. More information on produced and consumed tags can be found in the Rockwell publication *Logix5000 Controllers Produced and Consumed Tags* available at http://literature.rockwellautomation.com/idc/groups/literature/documents/pm/1756-pm011_-en-p.pdf.

Building Ladder Diagram programs (Simple)

Now we will start to create some PLC programs for our newly created project. In this recipe, we will learn how to create Ladder Logic programs in RSLogix 5000 by drafting a basic Ladder Logic alarm timer routine. We are going to add an alarm that will be triggered if the valve position set point and current analog input position of the valve differ for more than five seconds.

There are several editions of RSLogix 5000 available today, which are similar to Microsoft Windows' home and professional versions. The more "basic" (less expensive) editions of RSLogix 5000 have many features disabled. For example, only the full and professional editions, which are more expensive, support the editing of Function Block Diagrams, Graphical Structured Text, and Sequential Function Chart. In my experience, Ladder Logic is the most commonly used language. Refer to `http://www.rockwellautomation.com/rockwellsoftware/design/rslogix5000/orderinginfo.html` for more on this.

Getting ready

You will need to have added the cards and tags from the previous recipes to complete this exercise.

How to do it...

1. Open **Controller Organizer** and expand the leaf **Tasks | Main Tasks | Main Program**. Right-click on **Main Program** and select **New Routine** as shown in the following screenshot:

2. Configure a new Ladder Logic program by setting the following values:

    - **Name**: VALVES
    - **Description**: Valve Control Program
    - **Type**: **Ladder Diagram**

3. For our newly created routine to be executed with each scan of the PLC, we will need to add a reference to it in **MainRoutine** that is executed with each scan of the **MainTask** task.

4. Double-click on our **MainRoutine** program to display the Ladder Logic contained within it.

5. Next, we will add a **Jump To Subroutine (JSR)** element that will add our newly added Ladder Diagram program to the main task and ensure that it is executed with each scan.

6. Above the Ladder Diagram, there are tab buttons that organize **Ladder Elements** into **Element Groups**. Click on the left and right arrows that are on the left side of **Element Groups** and find the one labeled **Program Control**. After clicking on the **Program Control** element group, you will see the **JSR** element. Click on the **JSR** element to add it to the current **Ladder Logic Rung** in **MainRoutine**.

7. Next, we will make some modifications to the **JSR** routine so that it calls our newly added Ladder Diagram. Click on the **Routine Name** parameter of the JSR element and select the **VALVES** routine from the list as shown in the following screenshot:

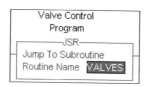

8. There are three additional parameters that we are not using as part of the JSR element, which can be removed. Select the **Input Par** parameter and then click on the Remove Parameter icon in the toolbar above the Ladder Diagram. This icon looks as shown in the following screenshot:

9. Repeat this process for the other optional parameter: **Return Par**.

10. Now that we have ensured that our newly added **Ladder Logic** routine will be scanned, we can add the elements to our **Ladder Logic** routine. Double-click on our **VALVES** routine in the **Controller Organizer** tab under the **MainTask** task.

11. Find the **Timer/Counter** element group and click on the **TON** (Timer On Delay) element to add it to our **Ladder Diagram**.

12. Now we will create the Timer object. Enter the name in the **Timer** field as FC1001_ TON. Right-click on the **TIMER** object tag name we just entered and select **New "FC1001_TON"** (or press *Ctrl + W*).

13. In the **New Tag** form that appears, enter in the description FAULT TIMER FOR FLOW CONTROL VALVE 1001 and click on **OK** to create the new **TIMER** tag.

14. Next, we will configure our **TON** element to count to five seconds (5,000 milliseconds). Double-click on the **Preset** parameter and enter in the value 5000, which is in milliseconds.

15. Now, we will need to add the condition that will start the **TIMER** object. We will be adding a Less Than (**LES**) element from the **Compare** element group. Be sure to add the element to the same Ladder Logic Rung as the **Timer on Delay** element.

16. The **LES** element will compare the valve position with the valve set point and return true if the values do not match. So set the two parameters of the **LES** element to the following:

   ❏ FC1001_PV

   ❏ FC1001_SP

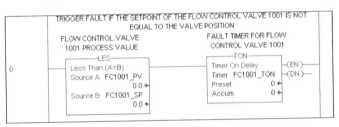

17. Now, we will add a second Ladder Logic Rung where a latched fault alarm is triggered after **TIMER** reaches five seconds.

18. Right-click under the first Ladder Logic Rung and select **Add Rung** (or press *Ctrl + R*).

19. Find the **Favorites** element group and select the Examine On icon as shown in the following screenshot:

20. Click on **?** above the **Examine On** tab and select the **TIMER** object's **Done** property, **FC1001_TON.DN**, as shown in the following screenshot. Now, once the valve values are not equal, and the **TIMER** has completed its count to five seconds, this Ladder Logic Rung will be activated as shown in the following screenshot:

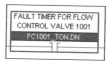

21. Next, we will add an **Output Latched** element to this Ladder Logic Rung. Click on the **Output Latched** element from the **Favorites** element group with our new rung selected.

22. Click on **?** above the **Output Latched** element and type in the name of a new base tag we are going to add as `FC1001_FLT`. Press *Enter* or click on the element to complete the text entry.

23. Right-click on **FC1001_FLT** and select **New "FC1001_FLT"** (or press *Ctrl + W*).

24. Set the following values in the **New Tag** form that appears:

   ❑ **Description**: FLOW CONTROL VALVE 1001 POSITION FAULT

   ❑ **Type**: Base

   ❑ **Scope**: FirstController

   ❑ **Data Type**: Bool

25. Click on **OK** to add the new tag. Our new tag will look like the following screenshot:

26. It is considered bad practice to latch a bit without having the code to unlatch the bit directly below it. Create a new **BOOL** type tag called `ALARM_RESET` with the following properties:

   ❑ **Name**: ALARM_RESET

   ❑ **Description**: RESET ALARMS

   ❑ **Type**: Base

   ❑ **Scope**: FirstController

   ❑ **Data Type**: BOOL

27. Click on **OK** to add the new tag. Then add the following coil and OTU to unlatch the fault when the master alarm reset is triggered.

28. Finally, we will add a comment so that we can see what our **Ladder Diagram** is doing at a glance.

29. Right-click in the far-right area of the first **Ladder Logic Rung** (where the **0** is) and select **Edit Rung Comment** (*Ctrl + D*).

30. Enter the following helpful comment:

```
TRIGGER FAULT IF THE SETPOINT OF THE FLOW CONTROL VALVE
1001 IS NOT EQUAL TO THE VALVE POSITION
```

How it works...

We have created our first Ladder Logic Diagram and linked it to the **MainTask** task. Now, each time that the task is scanned (executed), our Ladder Logic routine will be run from left to right and top to bottom.

There's more...

More information on Ladder Logic can be found in the Rockwell publication *Logix5000 Controllers Ladder Diagram* available at `http://literature.rockwellautomation.com/idc/groups/literature/documents/pm/1756-pm008_-en-p.pdf`.

Ladder Logic is the most commonly used programming language in RSLogix 5000. This recipe describes a few more helpful hints to get you started.

Understanding Ladder Rung statuses

Did you notice the vertical output **eeeeeee** on the left-hand side of your **Ladder Logic Rung**? This indicates that an error is present in your Ladder Logic code. After making changes to your controller project, it is a good practice to **Verify** your project using the drop-down menu item **Logic | Verify | Controller**. Once **Verify** has been run, you will see the error pane appear with any errors that it has detected.

Element help

You can easily get detailed documentation on Ladder Logic Elements, Function Block Diagram Elements, Structured Text Code, and other element types by selecting the object and pressing *F1*.

Copying and pasting Ladder Logic

Ladder Logic Rungs and elements can be copied and pasted within your ladder routine. Simply select the rung or element you wish to copy and press *Ctrl + C*. Then, to paste the rung or element, select the location where you would like to paste it and press *Ctrl + V*.

Troubleshooting techniques (Intermediate)

The ability to troubleshoot and properly diagnose faults in Logix 5000 is extremely valuable. This recipe will demonstrate a typical troubleshooting process when things go awry.

Getting ready

This recipe will cover the fundamentals of troubleshooting issues in RSLogix 5000. When there are troubleshooting issues in RSLogix 5000, there are a few places where you can start.

How to do it...

1. Open **Controller Organizer** in RSLogix 5000 and find the **Controller Module** (or modules for your project). Right-click on the **Controller** module and select **Properties** (or press *Alt + Enter*).

2. In the **Controller Properties** window, navigate to the **Major Faults** tab and check for any faults that may have occurred.

3. Now, click on the **Minor Faults** tab and check for any faults that may have occurred.

4. You can also check to see if there are any errors with the PLC program itself by navigating to **Logic | Verify | Controller**.

5. If any errors occur, they will be listed in the **Errors Pane** that is displayed.

How it works...

RSLogix 5000 allows you to resolve errors in your PLC program quickly and easily. Often, you can click on the error messages that appear and you will be taken directly to the area you need to fix.

There's more...

► More information on troubleshooting faults in RSLogix 5000 can be found in the Rockwell publication *Logix5000 Controllers Major, Minor, and I/O Faults* available at `http://literature.rockwellautomation.com/idc/groups/literature/documents/pm/1756-pm014_-en-p.pdf`

► And in the Rockwell publication *Logix5000 Controllers Information and Status* available at `http://literature.rockwellautomation.com/idc/groups/literature/documents/pm/1756-pm015_-en-p.pdf`

Downloading explained (Simple)

In this recipe, we will demonstrate how to download your application to a PLC in a step-by-step process.

Getting ready

To complete this recipe, you will need a PLC or RSLogix Emulate 5000 running.

How to do it...

1. Before we can communicate with our PLC, we will need to set up the communications path in a separate program called RSLinx that is installed with RSLogix 5000. Open **RSLinx** and then, from the drop-down menu, navigate to **Communications | Configure Drivers** as shown in the following screenshot:

2. The **Configure Drivers** window allows you to add **Driver** to the **Configured Drivers** list. In the **Available Driver Types** group box, use the dropdown and select the **EtherNet/ IP Driver** option, and then click on the **Add New...** button.

3.   Next, a window will appear asking you to specify the name of **Driver**; use the default driver name and click on **OK**.

4.   A **Configure Driver** window will appear next; simply click on **OK** to accept the default settings.

5.   We have completed our communications driver configuration in RSLinx, so we can close that application for now and switch back to RSLogix 5000.

6.   In order to download our program to the controller, we first need to specify the communications path. You can see the currently selected communications **Path** on **Communications Toolbar**:

7.   Click on the Who Active icon to the right-hand side of **Communications Toolbar** and specify the communications path. This icon looks like the following screenshot:

8.   The **Who Active** window will appear and allow you to specify the path to your PLC. Expand the **AB_ETHIP** driver we added using **RSLinx**, select your PLC, and then click on the **Download** button.

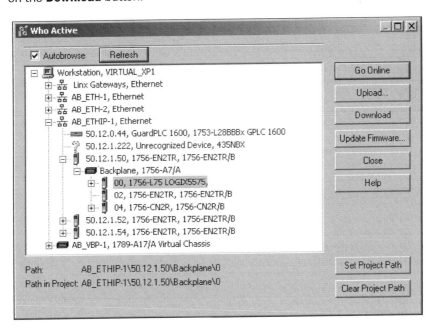

9. A version mismatch message may appear; just click on **OK** to accept and overwrite the current PLC program.

10. Now that we have downloaded the PLC program to the PLC, if you check the PLC status pane in the top-left corner of Logix 5000, you will see that you are now online.

Warning

Downloading a program to a PLC can be dangerous on a running plant. Downloading a new program or a new version of a program will stop a process that is running. Also, it is hard to be sure what code and values you are loading onto the PLC, so, when the process starts up again, things may go a bit haywire. If the plant simply cannot go down, you can make changes online (which we will explore in more detail later in the book).

Uploading explained (Simple)

In this recipe, we will demonstrate how to upload an application from a PLC to RSLogix 5000 in a step-by-step process.

Getting ready

To complete this recipe, you will need to have completed the previous recipe to configure your RSLinx to communicate with the device you need.

How to do it...

1. Click on the Who Active icon as shown in the following screenshot:

2. The **Who Active** window will appear. Use the network browser tree to select your PLC on the network.

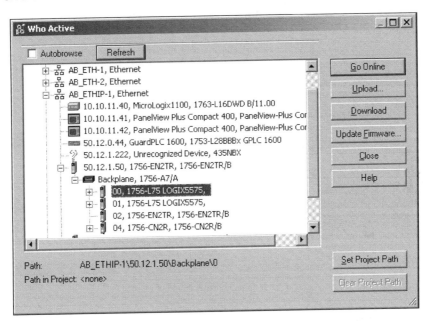

3. Now we are ready to upload the PLC program to our computer. Click on the **Upload** button.

4. The **Connected To Upload** window provides more details about the current PLC and the program state. Hit the **Upload** button to start the upload process.

5. If you have made any changes to the project that you currently have open, you will see a **Save Changes** dialog. Click on **No**.

6. An uploading progress bar will be displayed when the upload process begins.

7. Once the upload process is complete, you will see that you are now online on the PLC.

How it works...

Now that we have uploaded the current program from the PLC to our local copy of RSLogix 5000, the PLC program and our local program are an exact match.

There's more...

It is necessary to understand that comments are not stored in the PLC program that is uploaded to the controller.

Don't lose your comments and descriptions

One potential issue you can encounter when uploading a program from a PLC is that comments are not stored on the PLC. So, it is important to keep the following in mind when uploading programs on RSLogix 5000:

▶ If you upload a project from a PLC to a new project, the newly created project will be missing all comments and descriptions

▶ If someone adds new tags, descriptions, and comments to a program and downloads it to a PLC and you upload it to your local program that does not have the newly added tags, descriptions, and comments, your project will be missing the recently added descriptions and comments

Understanding online changes (Intermediate)

In this recipe, we will introduce the concept of **Online Changes** in RSLogix 5000. We are going to make an online modification to our **VALVES** routine by adding logic to check for a power-failure event.

Getting ready

In order to complete this recipe, you will need to have a PLC project that is free from errors, has been downloaded to the PLC, and is **Online** and **Equal**.

How to do it...

1. When you are online on your PLC (if you are not sure how to do this read the recipes *Uploading explained (Simple)* and *Downloading explained (Simple)*), right-click on **Program Tags**, and select **New Tag...** (or press *Ctrl + W*).

2. On the **New Tag** form that appears, create a new base type tag using the following values:

   ❏ **Name**: POWER_FAILURE

   ❏ **Description**: POWER FAILURE IN PROGRESS

   ❏ **Type: Base**

   ❏ **Data Type: BOOL**

   ❏ **Scope: MainProgram**

3. Now, we are going to add our newly created **POWER_FAILURE** tag to the **VALVES** Ladder Logic Routine. Open the **VALVES** routine by double-clicking on it from the **Controller Organizer** tree.

4. Now, the tricky part with an online change in RSLogix 5000 is that we cannot make modifications to an existing Ladder Logic that is currently running. However, we can add new Ladder Logic Rungs. So the workaround for editing rung is to copy and paste an existing rung, modify it, and then delete the old rung. So, let's copy the existing rung at position one by right-clicking on it and selecting **Copy** (or press *Ctrl + C*).

5. Next, we will paste our rung under the existing rung by right-clicking on it and selecting **Paste** (or press *Ctrl + V*).

6. After pasting the rung, you will notice that **iiiii** will appear to the left of the rung indicating that it is pending insertion.

7. Now that we have added the new rung, we can edit and go ahead and delete the old rung. Right-click on the only rung and click on **Delete** (or press *Delete*).

8. After deleting the old rung, you will notice that **ddddd** appears to the left of the rung indicating that it is pending deletion.

9. Now we can make our modification to our newly inserted rung. Choose the rung we have inserted and click on the **Examine Off** element to add it.

10. Click on **?** above the **Examine Off** element and associate it with the **POWER_FAILURE** tag. The following screenshot displays the **Examine Off** element with the associated tag.

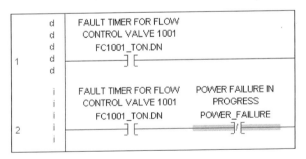

11. The changes we have made to the routine will not execute until we finalize our program (or accept, test, and assemble it). From the drop-down menu, navigate to **Logic | Online Edits | Finalize All Edits in Program** (or press *Ctrl + Shift + F*).

12. A dialog box will appear confirming that you want to finalize all edits to the program. Click on **Yes**.

13. After finalizing the edits, you will notice that the **iiiii** and **ddddd** indicators have disappeared and the new logic is now live.

How it works...

Online changes in RSLogix 5000 allow changes to be made to a running PLC without interrupting the code execution or the process it is controlling. In the later versions of RSLogix 5000, it is also possible to make online changes to FBD, ST, and SFC.

Building Functional Block Diagrams (Simple)

In this recipe, we will demonstrate how to use powerful **Functional Block Diagrams** (**FBD**) in your PLC program. We will create a simple digital alarm block routine using FBDs to manage our valve fault condition **FC1001_FLT**.

Getting ready

To complete this recipe, you should have completed the previous exercises.

How to do it...

1. Open the **Controller Organizer** window and expand the tree **Tasks | Main Tasks | Main Program**. Right-click and select **New Routine**.

2. Configure a new FBD routine by setting the following values:

    ❑ **Name**: DIGITAL_ALARMS

    ❑ **Description**: Digital Alarms

    ❑ **Type: Function Block Diagrams**

3. In order that our newly created routine executes with each scan of the PLC, we will need to add a reference to it in **MainRoutine**, which is executed with each scan of the **MainTask**. Repeat Steps 3 to 7 of the *Building Ladder Diagram programs* recipe (or copy and paste the existing JSR), but use the **Routine** name DIGITAL_ALARMS in the JSR element.

4. Now we will return to our DIGITAL_ALARMS FBD by double-clicking on it in the **Controller Organizer** window.

5. Next, we are going to add our **Digital Alarm** FBD, which we will use to manage our valve alarm fault. Select the **Alarms** element group in the FBD toolbox just above the FBD and click on the **ALMD** (Alarm Digital). The following screenshot displays the **Alarms** element group.

6. We need to connect the **ALMD** block to our valve fault alarm using an **Input Reference**, so let's add one to our FBD. The **Input Reference** object looks like an arrow (with a square corner) that is pointing to the right, as shown in the following screenshot. It can be found at the top-left area of the **Element Group** selector above the FBD. Click on the Input Reference object icon to add it to the diagram. This icon looks as shown in the following screenshot:

7. We will now set the input reference to point to our fault Base Tag. Click on the Input Reference icon, click on the drop-down option, and select **FC1001_FLT**.

8. Now we will need to reposition our blocks so that they fit properly on our FBD sheet. Click-and-drag the ALMD object to the right by a few inches.

9. Now we will connect the **FC1001_FLT** Input Reference to the **ALMD** block. Click and drag the point of the Input Reference object (you will see the mouse pointer change to a connector mouse icon) and release the mouse button over the Input Digital Pin object.

10. The ALMD function block that we added was automatically created as a base type object in our program scope tag list (**Program Tags**). We will now change the name of the ALMD object to follow our existing tag naming convention. Right-click on the top title of the ALMD object and navigate to the **Edit | ALMD_01** element.

11. Select the **Name** field in the element property and change it to: FC1001_FLT_ALM and click on **OK**. The scope of our FBD Base tag is set to MainProgram automatically when we added it to our routine. The routine now looks like the following screenshot:

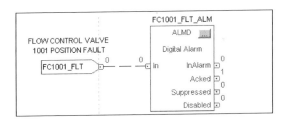

It is important to note that a Tag's scope cannot be changed after it has been created. You would need to delete and recreate the object in order to move it to another program or to the controller scope (global) level. If you wanted to create the function block at the controller scope level (a more global scope), you would need to declare the **ALMD** type tag manually at the controller scope level (using the **New Tag** form) and then change the FBD block to point to the **ALMD** tag you created.

How it works...

FBD routines are very different from Ladder Logic Routines; FBD routines are executed like a flow chart from input to output. FBD logic can flow in multiple directions and paths depending on how it is laid out. One or two FBD objects can easily replace dozens of Ladder Logic Rungs, thus making your program easier to maintain.

There's more...

More information on FBD can be found in the Rockwell publication *Logix5000 Controllers Function Block Diagram* available at `http://literature.rockwellautomation.com/idc/groups/literature/documents/pm/1756-pm009_-en-p.pdf`.

FBD provides a powerful set of high-level functions; it is important to understand how they interface with the HMI and how they are organized.

FBD properties

Double-clicking on an FBD block will open its properties. Each FBD block contains a unique set of properties and detailed help documentation is provided (by pressing the *F1* key). Many of the properties allow you to more tightly integrate your PLC controller with your **Human Machine Interface** (**HMI**) computer. Using FBD can allow you to configure many properties like alarm names in the PLC rather than in the HMI. Many SCADA system vendors are moving to a more DCS-style, single database configuration. Rockwell Automation's **PlantPAx** automation system takes this type of DCS functionality to the next level, but that is a subject for a separate book perhaps.

Organizing your FBD with sheets

You can also organize your FBDs into multiple sheets and use wire blocks to pass data between them. You can add sheets to your FBD by clicking on the New Sheet icon above your FBD routine. You can also provide a helpful name for each sheet by editing the **Sheet** text field.

The layout flexibility and Sheet organization that FBD routines provide make them more suitable for printing than Ladder Logic.

Building a Structured Text program (Simple)

In this recipe, we will explain a typical-use case for **Structured Text** (**ST**) programming. In our exercise, we will develop a simple Structured Text program for calculating the flow rate based on our valve position.

Getting ready

To complete this recipe you should have completed the previous exercises.

How to do it...

1. First, we will need to declare our new routine. Right-click on **MainProgram** of the **Controller Organizer** and select **New Routine**.

2. In the **New Routine** form that appears, enter or select the following:

    ❑ **Name**: CALCULATIONS

    ❑ **Description**: Process Calculations

    ❑ **Type**: **Structured Text**

3. Then click on **OK**.

4. In order for our newly created routine to be executed with each scan of the PLC, we will need to add a reference to it in **MainRoutine** that is executed with each Scan of the **MainTask**. Repeat Steps 3 to 7 of the *Building Ladder Diagram programs* recipe (or copy and paste the existing JSR), but use the routine name CALCULATIONS in the JSR element.

5. We will use a simple linear equation to calculate the `Flow` value based on our **Valve** position. Enter in the following structured text comment and simple formula code as shown in the following screenshot:

```
//Linear flow calculation based on valve position
FC1001_FLOW := FC1001_PV*0.83823;
```

```
//Linear flow calculation based on valve position
FC1001_FLOW := FC1001_PV*0.83823;
```

6. The tag `FC1001_FLOW` must be added to our controller, so right-click on the `FC1001_FLOW` tag in our structured text and select **New Tag "FC1001_FLOW"**.

7. On the **New Tag** form, enter the following values:

    ❏ **Name**: `FC1001_FLOW`

    ❏ **Description**: `Flow estimated based on valve position`

    ❏ **Type**: **REAL**

    ❏ **Scope**: **FirstController**

8. Then click on **OK**.

How it works...

Structured Text (ST) is similar to traditional programming languages, such as Pascal, C, and BASIC, except that it will run continuously from start to finish as if it were contained within a loop. Structured Text is a great place to put complex formulas that would be difficult to implement using Ladder Logic.

There's more...

More information on ST can be found in the Rockwell publication *Logix5000 Controllers Structured Text* available at `http://literature.rockwellautomation.com/idc/groups/literature/documents/pm/1756-pm007_-en-p.pdf`.

Building a Sequential Function Chart (Intermediate)

In the following recipe, we will demonstrate the usage of a **Sequential Function Chart (SFC)** by building a backwash process using a step-by-step guide.

Getting ready

To complete this recipe, you should have completed the previous exercises.

How to do it...

1. First, we will need to declare our new routine. Right-click on **MainProgram** of **Controller Organizer** and select **New Routine**.

2. In the **New Routine** form that appears, enter or select the following:

   ❑ **Name**: BACKWASH

   ❑ **Description**: Backwash Sequence

   ❑ **Type: Sequential Function Chart**

3. Then click on **OK**.

4. In order for our newly created routine to be executed with each scan of the PLC, we will need to add a reference to it in **MainRoutine** that is executed with each Scan of **MainTask**. Repeat Steps 3 to 7 of the *Building Ladder Diagram programs* recipe (or copy and paste the existing JSR), but use the routine name BACKWASH in the JSR element.

5. Now we will add an Action to Step_000 that was created by default. Right-click on Step_000 and select **Add Action**.

6. Now we will add the initialization values for our SFC using ST syntax. Double-click on the box with **?** at the bottom of **Action_000** and enter the following structured text as shown in the following screenshot:

```
BACKWASH_START_PB:=0;
BACKWASH_FLT:=0;
```

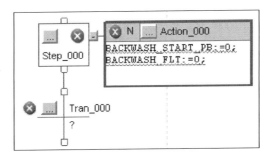

7. An icon with a red cross will appear to the side of **Action_000**, indicating that there is an error with the ST we have added. The error is due to the fact that BACKWASH_START and BACKWASH_FLT do not yet exist in our program.

8. We can easily add these values by right-clicking on the BACKWASH_START_PB tag in the Action_000 box and selecting the **New Tag "BACKWASH_START_PB"** option.

9. The **New Tag** form will appear, which will allow us to create our new tag as follows:

   ❑ **Name**: BACKWASH_START_PB

   ❑ **Description**: START BACKWASH PUSH BUTTON

   ❑ **Type: BOOL**

   ❑ **Scope: FirstController**

10. Repeat the same process for the second new tag by right-clicking on the tag BACKWASH_FLT and the **New Tag** form and entering the following values:

    ❑ **Name**: BACKWASH_FLT

    ❑ **Description**: BACKWASH SEQUENCE FAULT

    ❑ **Type: BOOL**

    ❑ **Scope: FirstController**

11. Next, we will add the Transition conditional value that will start our Backwash sequence. Double-click on the Tran_000 question mark and enter the following logical statement (which is equivalent to BACKWASH_START_PB=1) BACKWASH_START_PB.

12. Now we can add our backwash sequence step. Select the Transition box **Tran_000** and then click on the **Step** element icon in the **SFC** element group above our **SFC** program. The step box, Step_001, will be added and automatically connected to Tran_000 (because we selected it before adding our new step).

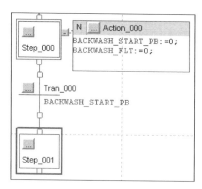

13. Next, we can add an action to `Step_001` that will represent our backwash process. Right-click on `Step_001` and select **Add Action**.

14. Now we will add our action ST to the action element by double-clicking on **?** in `Action_001` and entering `FC1001_SP:=100;`.

15. Let's add a delay to `Step_001` in order to give our backwash time to complete. Right-click on **Step_001** and select the **Step Properties** menu option. On the **Step Properties** form, set the **Preset** field to `30000` ms (30 seconds). Click on **OK**.

16. Next, we will add a Selection Branch Diverge in order to reset our sequence or trigger a fault if there is a problem. Select `Step_001` and then click on the Selection Branch Diverge Element icon just above our sequence chart. The icon is shown in the following screenshot:

17. A Selection Branch Diverge can also allow you to execute one or another sequence, while Simultaneous Branch Diverge will execute two sequences in parallel. Our sequence will automatically reset and await another backwash if the valve has successfully opened, 100 percent. Select **Tran_001** and click on the **?** icon to set the logical statement that will execute this Selection Branch. Type in the following structured text logical statement: `FC1001_PV=100`.

18. We want our sequence to reset after it has completed the backwash, so we will connect a flow line from our transition `Tran_001` to the top element in our sequence, `Step_001`. Click on the connector box under `Trans_001` and drag it to the connector box on top of **Step_000**.

19. If our valve fails to open, we want to raise a fault before resetting our sequence. Select **Tran_002**, click on **?**, and enter the following structured text logical statement; `FC1001_PV<>100`.

20. In order to raise a fault, we will need to add a Step. Select `Tran_002` and click on the Step Element icon.

21. Add an action to our newly created step, `Step_002`, by right-clicking and selecting **Add Action**.

22. Double-click on **?** of our newly added action and add the following ST code:

```
BACKWASH_FLT:=1;
BACKWASH_START_PB:=0;
```

23. In order to make our sequence easy to understand, let's add a Text Box comment. Click on the Text Box element to add it to the sequence diagram and drag it to the right of the sequence. Enter in the following comment: `Backwash fault triggered if valve FC1001 fails to open.`

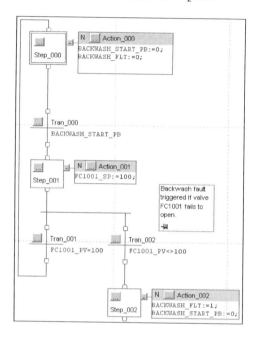

How it works...

The previous programs we have developed are ideal for a single state whereas SFC routines are well suited for more complex processes that require simultaneous operations. SFCs are steps that are connected to transitions. A flow line will connect the step and transition in the sequence. Transitions are the triggers between each step and each step acts as a collection of actions. A Selection Branch Diverge will follow one or the other path and a Simultaneous Branch Diverge will execute two sequences at the same time.

There's more...

More information on SFC can be found in the Rockwell publication *Logix5000 Controllers Sequential Function Charts* available at `http://literature.rockwellautomation.com/idc/groups/literature/documents/pm/1756-pm006_-en-p.pdf`. In the interest of saving time in the last exercise, I did not follow the best practice of providing useful names to the steps and transitions.

Renaming sequence steps and transitions

In the previous exercise, we left the default names for our steps, actions, and transitions as they were. It is easy to rename the SFC elements to make the sequence easier to read and maintain. Simply double-click on the SFC name to rename it.

Organizing your project code (Advanced)

In this recipe, we will explore the way RSLogix organizes its controller into Scope, Tasks, Programs, and Routines.

Getting ready

To complete this recipe, you should have completed the previous exercises.

How to do it...

1. In **Controller Organizer**, right-click on the Tasks icon and click on **New Task**.

2. In our project, there is no need to check for our non-critical alarms every 10 ms. We will create a new periodic task for processing alarms every 250 ms and give it low priority in order to reduce the load on our processor.

3. In the **New Task** form that appears, enter the following values:

   □ **Name**: AlarmTask

   □ **Description**: Task for Calculating Alarm Conditions

   □ **Type**: **Periodic**

   □ **Period**: 250.000 **ms**

   □ **Priority**: 11

   □ **Watch Dog**: 500 **ms**

4. Next, we will add our Program that will contain our Alarm function and handle the alarm processing for our project.

5. Right-click on our newly created AlarmTask task and click on the **New Program** option.

6. In the **New Program** form that appears, enter the following values:

    ❑ **Name**: MainAlarmProgram

    ❑ **Description**: Program for processing Alarms

7. Now we can move the **DIGITAL_ALARM** routine, which we created earlier in the book, to our newly created **Alarm** program. Expand **MainTask** and **MainProgram** in the **Controller Organizer** Tasks and drag-and-drop the **DIGITAL_ALARM** routine from **MainProgram** to **MainAlarmProgram** as shown in the following screenshot:

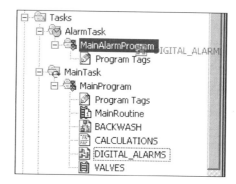

8. As we learned earlier, in order for a routine to be executed, it must be linked to a program. We will now set the **DIGITAL_ALARM** routine as the MainRoutine of **MainAlarmProgram**. Right-click on **MainAlarmProgram** and select **Properties** (or hit *Alt + Enter*). In the **Program Properties** form that appears, select the **Configuration** tab and, under the **Assign Routines** header, select **DIGITAL_ALARM** from the **Main** dropdown as shown in the following screenshot:

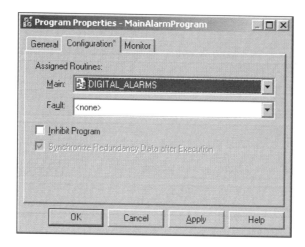

9. You will notice that, after assigning the **DIGITAL_ALARM** routine as the MainRoutine for **MainAlarmProgram** as shown in the following screenshot, the **DIGITAL_ALARM** icon changes to display **1** that indicates that it is the MainRoutine for the program:

10. Next, we will check to see whether or not we have introduced any errors in our controller with our latest changes. From the drop-down menu at the top of RSLogix 5000, navigate to **Logic | Verify | Controller**.

11. You will notice that the **Errors** pane has appeared and the following errors are listed:

    ❑ **Error: Sheet 1, B1, ALMD, FC1001_FLT_ALM: Tag doesn't reference valid object or target.**

    ❑ **Error: Rung 1, JSR, Operand 0: Invalid reference to unknown routine.**

12. Clicking on the first error message will directly take you to the **DIGITAL_ALARMS** FBD and highlight the **FC1001_FLT_ALM** ALMD element.

 The red cross mark on the FBD element indicates that there is a problem with the block's configuration.

13. The error has occurred because the original **FC1001_FLT_ALM** was created with a scope of **MainProgram** and it cannot be accessed from the **MainAlarmProgram**.

 FBD tags are automatically created with the scope of the current program you are working under when they are added to a diagram routine. In order to declare an FBD at the Controller scope (Global scope) level you will need to create it manually using the **New Tag** form.

14. In order to fix this problem, we will need to create the ALMD tag again at the **Controller** scope. Right-click on the **FC1001_FLT_ALM** tag and select the **New "FC1001_FLT_ALM"** menu option (or press *Ctrl + W*).

15. The **New Tag** form will appear. Ensure that the **Scope** field is set to **FirstController** and click on **OK**.

16. Next, we should remove the duplicate **FC1001_FLT_ALM** tag that exists in the **MainProgram** Scope. Under the **MainTask** and **MainProgram** folders of **Controller Organizer**, right-click on the **Program Tags** icon and select **Edit Tags**.

17. The **Edit Tags** table will appear; select the **FC1001_FLT_ALM** tag, right-click on the box to the left of the name, and select the **Delete** menu option (or press the *Delete* key).

18. We have now fixed the first error message; now, let's resolve the second. Clicking on the second error message in the error pane will take you directly to the **Jump To Subroutine (JSR)** reference within our **MainProgram** element's **MainRoutine** object to the **DIGITAL_ALARM** routine. The error is being displayed because the **DIGITAL_ALARM** routine is no longer in the **MainProgram** element's scope. Delete this Ladder Logic Rung by right-clicking on it and selecting **Delete** (or by pressing the *Delete* key while clicking on the rung selected).

19. Finally, we will **Verify** the program once more to ensure that we no longer have any problems.

20. From the drop-down menu at the top of RSLogix 5000, navigate to **Logic | Verify | Controller**.

How it works...

As you can see from the exercise, your routines can be linked to different Tasks, which allows you to control how and when they are executed, and that can reduce the processing load on your PLC.

There's more...

Tasks provide us with control over how frequently sections of code will execute or if they will execute at all. It is critical to understand that these tasks are executed asynchronously in the world of RSLogix 5000, so let's take a look at a few features of Tasks in more detail.

Inhibit programs and tasks

One advantage of dividing your project into Tasks and Programs is that you can inhibit them individually, if needed, and prevent them from executing. To inhibit a Task or Program, right-click on it and open its properties. The **Inhibit** option is on the **Configuration** tab of the **Properties** window.

Understanding task types

There are three types of tasks in RSLogix 5000: Periodic, Event, and Continuous. **Periodic tasks** will run at an interval that you can specify in milliseconds. **Event tasks** will run when a predefined condition triggers it. **Continuous tasks** will execute as quickly as they can. Only one continuous task is allowed to be declared per controller. The main task that is added by default when you create a new project is automatically set up as a continuous task. More information on task types can be found in the Rockwell publication *Logix5000 Controllers Tasks, Programs, and Routines* available at `http://literature.rockwellautomation.com/idc/groups/literature/documents/pm/1756-pm001_-en-e.pdf`.

Exporting tags (Simple)

Exporting programs and tags is a handy way to generate reports on your RSLogix 5000 application and create a handle tag list for integration with an HMI.

Getting ready

In order to complete this recipe, you need an RSLogix 5000 project to export.

How to do it...

1. Open your RSLogix 5000 project and navigate from the drop-down menu **Tools | Export | Tags and Logic Comments**.

2. The **Export** window will allow you to choose a location on your computer to save the CSV file that is produced. Select a location on your computer and click on **Save**.

3. The CSV file that was generated can be opened and viewed using MS Excel. Open the file using MS Excel and resize the columns to make the document easier to read.

How it works...

Exporting your project tags to a CSV file will allow you to share and manipulate the tags more easily.

There's more...

Often, when exporting tags from RSLogix 5000, your aim will be to modify them and re-import them into your program.

Importing programs

After exporting tags to a CSV file, you can also import them back into your project by navigating to the drop-down menu option **Tools | Import | Tags and Logic Comments**.

This makes it very easy to bulk-add tags using Excel or make changes to multiple tag descriptions.

 When importing tags, it is important to note that the import is "keyed" on the tag name. For example, if you make a change to a tag and do not change the tag name, the tag info will be updated. If you change a tag name and then import the tag list, that tag will be treated as if it were a newly added tag.

Exporting programs (Simple)

In this recipe, we will demonstrate how to export a program to an **XML**-based (**Extensible Mark-up Language**) **L5X** file for manipulation or reporting.

Getting ready

In order to complete this recipe, you will need an RSLogix 5000 program to export.

How to do it...

1. From RSLogix 5000 **Controller Organizer**, right-click on a program and select the **Export Routine...** option.

2. Use the **Export** window file browser to select a location on your computer to save your L5X file. Click on the **Save** button to store the **L5X** file on your computer.

3. Open the L5X on your computer using Notepad by right-clicking on the file and selecting **Open With...**; then select **Notepad** from the list.

4. View the XML file in **Notepad**.

How it works...

The L5X file is an XML-based representation of your program. The L5X file can be used to make widespread changes to your program, for reporting purposes, or for importing by third-party products. More information about importing and exporting programs from RSLogix 5000 can be found in the *Rockwell* publication *Logix5000 Controllers Import/Export Project Components* available at http://literature.rockwellautomation.com/idc/groups/literature/documents/pm/1756-pm019_-en-p.pdf.

There's more...

The L5X file can also be easily modified and re-imported into your project.

Importing an L5X

You can use the L5X file to modify your program and re-import it into your project. For example, if there are repeated parts of your plant, you can perform **Find / Replace** of the tags that will change between each area using Notepad. Once your changes are in place, you can re-import the L5X file by right-clicking on **Controller Organizer** and selecting the **Import Routine...** option, and then navigating to the L5X file you wish to import.

The **Import Configuration** window provides you with a wide range of options for importing your routine into your project as shown in the following screenshot:

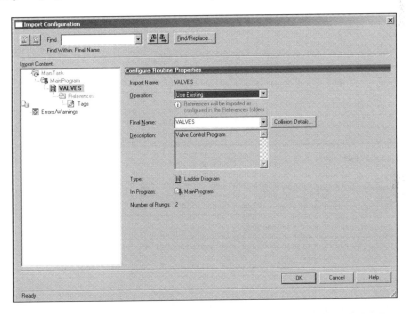

Printing programs (Simple)

In this recipe, we will demonstrate the best practices for printing Ladder Logic and Functional Block Diagrams for hard-copy documentation or engineering review meetings.

Getting ready

In order to complete this recipe, you will need to have a Ladder Logic Routine and a Function Block Diagram Routine to print.

How to do it...

1. Open a Ladder Logic Routine by double-clicking on it in **Controller Organizer** (you can use the **VALVES** routine that we created earlier in the book).

2. Open the drop-down menu option by navigating to **File | Print Options**. **Print Options** allow us to modify the data that is printed and how it is displayed for the wide range of reports we can generate from RSLogix 5000 Review the available items in the **Print Options** window.

3. Under the **Tag Listing** list box item, click to uncheck the **Expand Arrays** option.

4. Return to **Ladder Logic Routine** and, from the drop-down menu, navigate to **File | Print** (or press *Ctrl + P*).

5. Open a **Function Block Diagram** by double-clicking on it in **Controller Organizer** (you can use the DIGITAL_ALARM routine we created earlier in the book).

6. We want our FBD routine to print to a standard 11 x 17 sheet of paper so that it can be included alongside our P&ID drawings. Right-click on the background of the FBD routine and select **Properties** (or press *Alt + Enter*).

7. Click on the **Sheet Layout** tab and, from the **Sheet Size** drop-down selection, click on the option **Tabloid: 11 x 17 in**. Then click on the **OK** button. This will also provide a much larger area to lay out your FBD routines.

8. Print the FBD by opening the drop-down menu and navigating to **File | Print** (or press *Ctrl + P*).

How it works...

RSLogix offers a wide range of options for printing our tags, logic, and reports. Function Block Diagrams can benefit from a larger Sheet Size as it provides a larger area to lay out your routines.

Code generation (Advanced)

Often, there is a need to duplicate the same logic many times, but with different variable tags. This can be easily accomplished in Logix5000 using Notepad and a simple code generation trick (of the true Logix5000 masters).

Getting ready

In order to complete this recipe, you will need to have completed the previous recipe on Functional Block Diagrams.

How to do it...

1. First, we will open our existing Ladder Logic Routine using **Controller Organizer** and double-clicking on the **VALVES** routine.

2. Select both the valve rungs and copy them to the clipboard by clicking on the numbers **0** and **1** while holding *Ctrl* and then right-clicking on the rung and selecting **Copy Rung** (or pressing *Ctrl + C*).

3. Next, open Notepad and paste the following copied rungs onto it by selecting **Edit | Paste** (or by pressing *Ctrl + V*):

   ```
   NEQ(FC1001_PV,FC1001_SP)TON(FC1001_TON,?,?);
   XIC(FC1001_TON.DN)OTL(FC1001_FLT);
   ```

4. You will notice that the Ladder Logic has been copied in a programming language called **Instruction List** (**IL**), which is defined as part of the IEC 61131 standard. Now that the code is in plain text IL, we can edit and duplicate it more easily.

5. Now we will duplicate the code two more times for the remaining three valves we will add to our project. Copy and paste the following code two more times in Notepad:

   ```
   NEQ(FC1001_PV,FC1001_SP)TON(FC1001_TON,?,?);
   XIC(FC1001_TON.DN)OTL(FC1001_FLT);
   NEQ(FC1001_PV,FC1001_SP)TON(FC1001_TON,?,?);
   XIC(FC1001_TON.DN)OTL(FC1001_FLT);    NEQ(FC1001_PV,FC1001_SP)
   TON(FC1001_TON,?,?);
   XIC(FC1001_TON.DN)OTL(FC1001_FLT);
   ```

6. Next, we will modify the tag prefix used on each copy we've made. Each tag of the first IL copy should be changed from `FC1001` to `FC1002` and so on:

   ```
   NEQ(FC1002_PV,FC1002_SP)TON(FC1002_TON,?,?);
   XIC(FC1002_TON.DN)OTL(FC1002_FLT);
   NEQ(FC1003_PV,FC1003_SP)TON(FC1003_TON,?,?);
   XIC(FC1003_TON.DN)OTL(FC1003_FLT);
   NEQ(FC1004_PV,FC1004_SP)TON(FC1004_TON,?,?);
   XIC(FC1004_TON.DN)OTL(FC1004_FLT);
   ```

7. Next, we will copy all of our newly edited IL code in Notepad to our clipboard by selecting all of it and then navigating to the drop-down menu **Edit | Copy** (or by pressing *Ctrl + C*).

8. Returning to our RSLogix 5000 Ladder Logic Routine, we will now paste the IL code into our routine. Right-click on the last rung of our routine that has the **(end)** label and select **Edit Rung** (or press the *Enter* key). The Rung IL code editing text field will appear at the top of Ladder Logic.

9.   Select the entire contents of the Rung IL code in the text field and press the *Delete* key to clear it.

10.   Next, paste the code we've created in Notepad into the Rung edit IL code text field by right-clicking on it and selecting **Paste** (or by pressing *Ctrl + V*).

11.   Now we will add our updated IL code to our Ladder Logic Routine by pressing the green check mark button beside the IL code edit form field.

12.   We will need to add the missing tags and timer blocks before we can verify and download our application.

How it works...

Anytime we copy Ladder Logic to the clipboard, it is stored in both Ladder Logic (used internally by RSLogix 5000) and in the text-based IL programming language, and can be pasted into Notepad or other text-editing applications. IL is a low-level IEC standard language that all other IEC programming languages can be converted to. IL is most often used on old handheld programmers.

Understanding user-defined data types (Advanced)

User Defined Data Types (**UDTs**) are an extremely powerful feature in the RSLogix 5000 platform that improves the maintainability, uniformity, and readability of our routines. UDTs are groups of base data types (`BOOL`, `REAL`, `INT`, `TIMER`, and so on), which we can define and re-use.

Getting ready

In order to complete this recipe, you will need to have an understanding of RSLogix 5000 base data types.

How to do it...

1.   Expand the RSLogix 5000 **Controller Organizer** tab, find and expand the **Data Types** folder, and select the **User-Defined** node.

2.   Right-click on the **User-Defined** node and select **New Data Type...**.

3. The **Data Type** window that appears will allow us to configure our UDT. Enter the following name and description in the fields provided:

   ❑ **Name**: VALVE

   ❑ **Description**: Valve UDT to manage Valve Positions and Alarms

4. The data table under the **Description** UDT allows us to define the tags that will be included with each instance of the structure we've created. If we had created this UDT earlier, it would have saved us quite a bit of work creating all the tags for our various flow control valves (**FC1001**, **FC1002**, **FC1003**, and so on). Let's add our first tag to the table to replace our flow control valve position feedback variable FC1001_PV. Add the following values to the first row of the table:

   ❑ **Name**: PV

   ❑ **Data Type**: **REAL**

   ❑ **Description**: Current Valve Position

5. Next, we can enter the remaining simple base types to the table as follows:

   ❑ **SP**: **REAL**

   ❑ **FLOW**: **REAL**

   ❑ **FLT**: **BOOL**

6. UDTs also allow us to create more complex base types inside them, such as the **TIMER** type and even other UDTs. Let's add the complex base type tags to our UDT `TON: TIMER` as shown in the following screenshot:

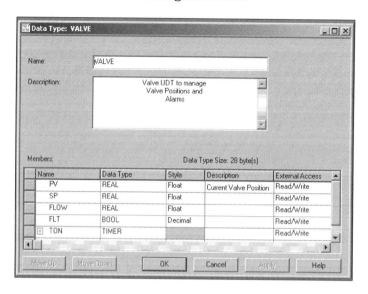

7. Click on **OK** to save our UDT. Now we have recreated all of our flow control valve tags in a single UDT.

8. Next, we will create an instance of our VALVE UDT and link it to our Analog In card channel. On **Controller Organizer**, right-click on **Controller Tags** under the **Controller** folder and select **New Tag...** (or press *Ctrl + W*).

9. In the **New Tag** window, enter the following values:

   ❑ **Name**: **FC1001**

   ❑ **Description**: Flow Control Valve 1001

   ❑ **Data Type**: **VALVE**

10. Click on the **OK** button to save the new tag.

11. Next, we will link one of the values of our UDT tag using Ladder Logic. Open the **VALVES** Ladder Logic Routine and add a new rung by right-clicking and selecting **Add Rung** (or press *Ctrl + R*).

12. Add a **Move** element to the new rung by selecting it under the **Move/Logical** element group.

13. Set the source of the element to `FC1001_PV` and set the destination of the element to `FC1001.PV` as shown in the following screenshot:

How it works...

One major advantage of UDTs is that they can be changed at any time (as of Version 17 and higher or RSLogix 5000, they can also be modified online); so, if you realize you forgot to add a tag to your UDT halfway through a project, you can easily make this modification.

There's more...

In this exercise, we barely scratched the surface of the power of UDTs; nested UDTs can be used to further improve code maintainability and consistency.

Nesting UDT

UDTs can be nested inside other UDTs as required. The nested UDT can be assigned just like any normal data type in RSLogix 5000.

FBD and SPC types not allowed

Many base types are available to be used in UDT structures; however, FBD and SPC cannot be added to UDT structures.

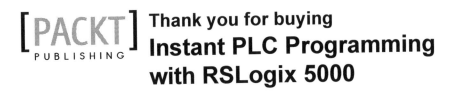

Thank you for buying
Instant PLC Programming with RSLogix 5000

About Packt Publishing

Packt, pronounced 'packed', published its first book "*Mastering phpMyAdmin for Effective MySQL Management*" in April 2004 and subsequently continued to specialize in publishing highly focused books on specific technologies and solutions.

Our books and publications share the experiences of your fellow IT professionals in adapting and customizing today's systems, applications, and frameworks. Our solution based books give you the knowledge and power to customize the software and technologies you're using to get the job done. Packt books are more specific and less general than the IT books you have seen in the past. Our unique business model allows us to bring you more focused information, giving you more of what you need to know, and less of what you don't.

Packt is a modern, yet unique publishing company, which focuses on producing quality, cutting-edge books for communities of developers, administrators, and newbies alike. For more information, please visit our website: www.packtpub.com.

Writing for Packt

We welcome all inquiries from people who are interested in authoring. Book proposals should be sent to author@packtpub.com. If your book idea is still at an early stage and you would like to discuss it first before writing a formal book proposal, contact us; one of our commissioning editors will get in touch with you.

We're not just looking for published authors; if you have strong technical skills but no writing experience, our experienced editors can help you develop a writing career, or simply get some additional reward for your expertise.

Multimedia Programming with Pure Data

ISBN: 978-1-78216-464-7   Paperback: 350 pages

A comprehensive guide for digital artists for creating rich interactive multimedia applications using Pure Data

1. Carefully organized topics for interactive multimedia professional practice

2. Detailed reference to a large collection of resources in the open source communities to enhance the Pure Data software

3. Visual explanation and step-by-step tutorials with practical and creative multimedia applications

Unity 4.x Game AI Programming

ISBN: 978-1-84969-340-0   Paperback: 232 pages

Learn and implement game AI in Unity3D with a lot of sample projects and next-generation techniques to use in your Unity3D projects

1. A practical guide with step-by-step instructions and example projects to learn Unity3D scripting

2. Learn pathfinding using A* algorithms as well as Unity3D pro features and navigation graphs

3. Implement finite state machines (FSMs), path following, and steering algorithms

Please check **www.PacktPub.com** for information on our titles

PUBLISHING

Unreal Development Kit Game
Programming with UnrealScript

Create games beyond your imagination with the Unreal
Development Kit

Beginner's Guide

Rachel Cordone

Unreal Development Kit Game Programming with UnrealScript: Beginner's Guide

ISBN: 978-1-84969-192-5   Paperback: 466 pages

Create games beyond your imagination with the
Unreal Development Kit

1. Dive into game programming with
   UnrealScript by creating a working
   example game.

2. Learn how the Unreal Development Kit is
   organized and how to quickly set up your
   own projects.

3. Recognize and fix crashes and other errors
   that come up during a game's development.

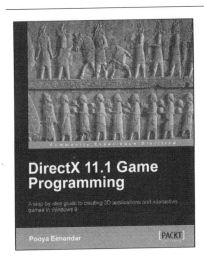

DirectX 11.1 Game
Programming

A step-by-step guide to creating 3D applications and interactive
games in Windows 8

Pooya Eimandar

DirectX 11.1 Game Programming

ISBN: 978-1-84969-480-3   Paperback: 240 pages

A step-by-step guide to creating 3D applications and
interactive games in Windows 8

1. Learn new features in Direct3D 11.1

2. Discover how to develop a multithreaded
   pipeline game engine

3. Understand shader model 5 and learn how to
   create an editor for the game

Please check **www.PacktPub.com** for information on our titles

Made in the USA
Lexington, KY
04 July 2014